Marshmallows, Mud Pies, and
MESSY ROOMS

Rebecca Woodbury, Ph.D., M.Ed.

Gravitas Publications Inc.

Marshmallows, Mud Pies, and
MESSY ROOMS

Illustrations: Janet Moneymaker

Marshmallows, Mud Pies, and Messy Rooms
ISBN 978-1-950415-13-7

Published by Gravitas Publications Inc.
Imprint: Real Science-4-Kids
www.gravitaspublications.com
www.realscience4kids.com

RS4K

Photo credits: Cover, Title Pg, P.2. : By MNStudio, AdobeStock; Above, By Zarina Lukash, AdobeStock; P.3. By Africa Studio, AdobeStock; P.5. By Zarina Lukash, AdobeStock; P.7. By GiltonF from Pixabay; P.13. By Nandalal, AdobeStock; P.20. Top, By bit24, AdobeStock; Middle, T-shirt, By airdone, AdobeStock & Sweater, By nys, AdobeStock; Bottom, By P, AdobeStock; P.21. Medicine & Watercolors, By SeventyFour, AdobeStock; Gas pump, By manusapon, AdobeStock; Brushing, By didesign, AdobeStock

What happens when you
pour water into dirt?

Look!
A mud pie.

What happens when you mix

eggs, flour, lemon, and water?

Yummy!
A lemon pie.

Mud pies and lemon pies are **mixtures**.

A **mixture** is any two or more things mixed together.

Mixtures can be made of large things like toys and books.

Is cheese a mixture?

Mixtures can be made of small things like **water molecules** and **salt molecules.**

We are small.

Are we a mixture?

Salt Water in a Glass

Water molecule

Salt molecule

Review: ATOMS

- **Atoms** are tiny building blocks that can link together.

- **Atoms** make everything we see, touch, taste, and smell.

Review: MOLECULES

Molecules are made when **atoms link** together.

Sometimes mixtures need to be separated.

Some mixtures
can be separated
by hand.

Some mixtures
can be separated
using **tools.**

Sometimes mixtures can be separated using tricks!

Paper chromatography is a trick used to separate the colors in ink.

Paper Chromatography

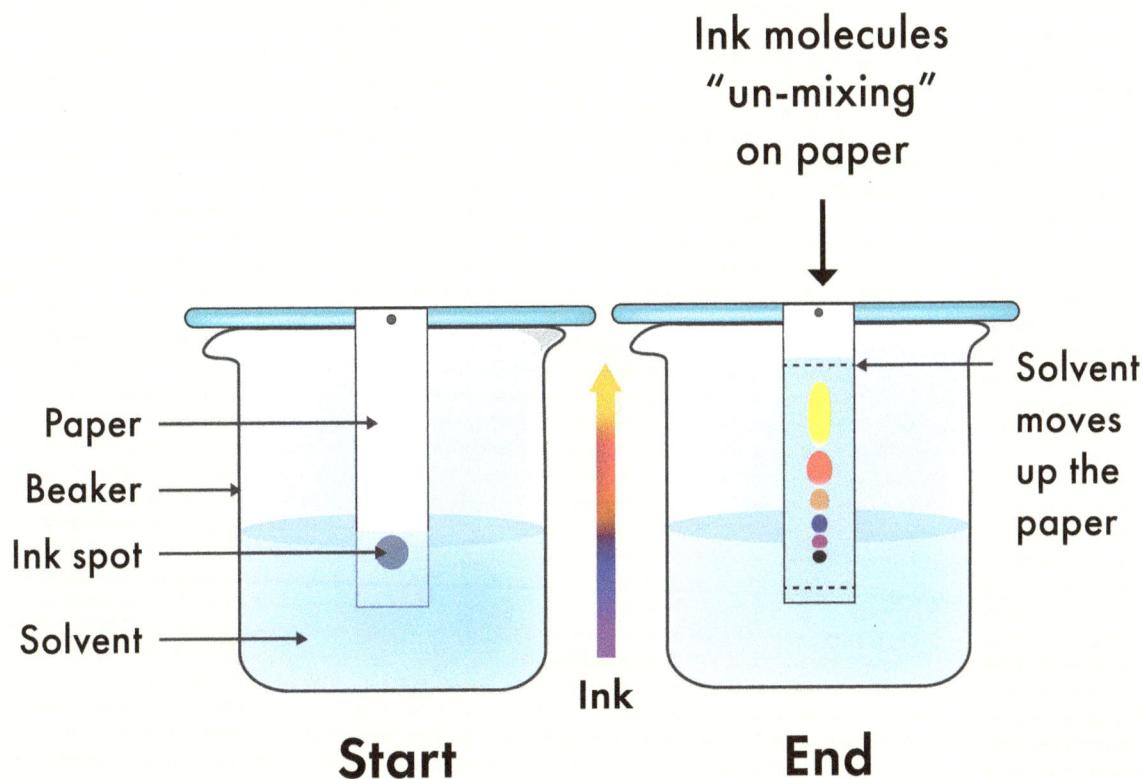

Ink molecules "un-mixing" on paper

Paper

Beaker

Ink spot

Solvent

Ink

Solvent moves up the paper

Start

End

Ink that is a mixture of different colors can be separated with the use of chromatography paper.

Filtration is a trick used to separate large particles from small particles.

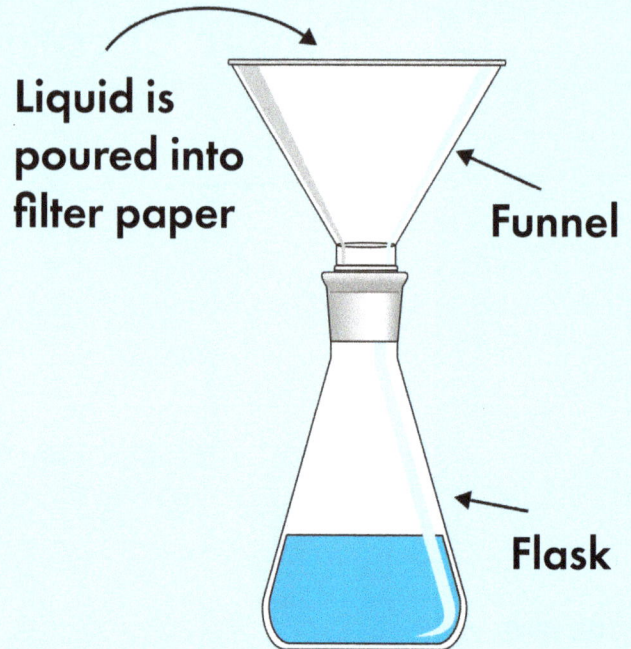

Liquid is poured into filter paper

Funnel

Flask

Filtration can be used to separate shrimp from oil.

Evaporation is a trick used by chefs to separate water or alcohol from solid foods.

Yo ho ho and a bottle of rum.

Soap molecules use a trick to separate oil molecules from water molecules. Soap molecules have a water part and an oily part.

Soap Molecule

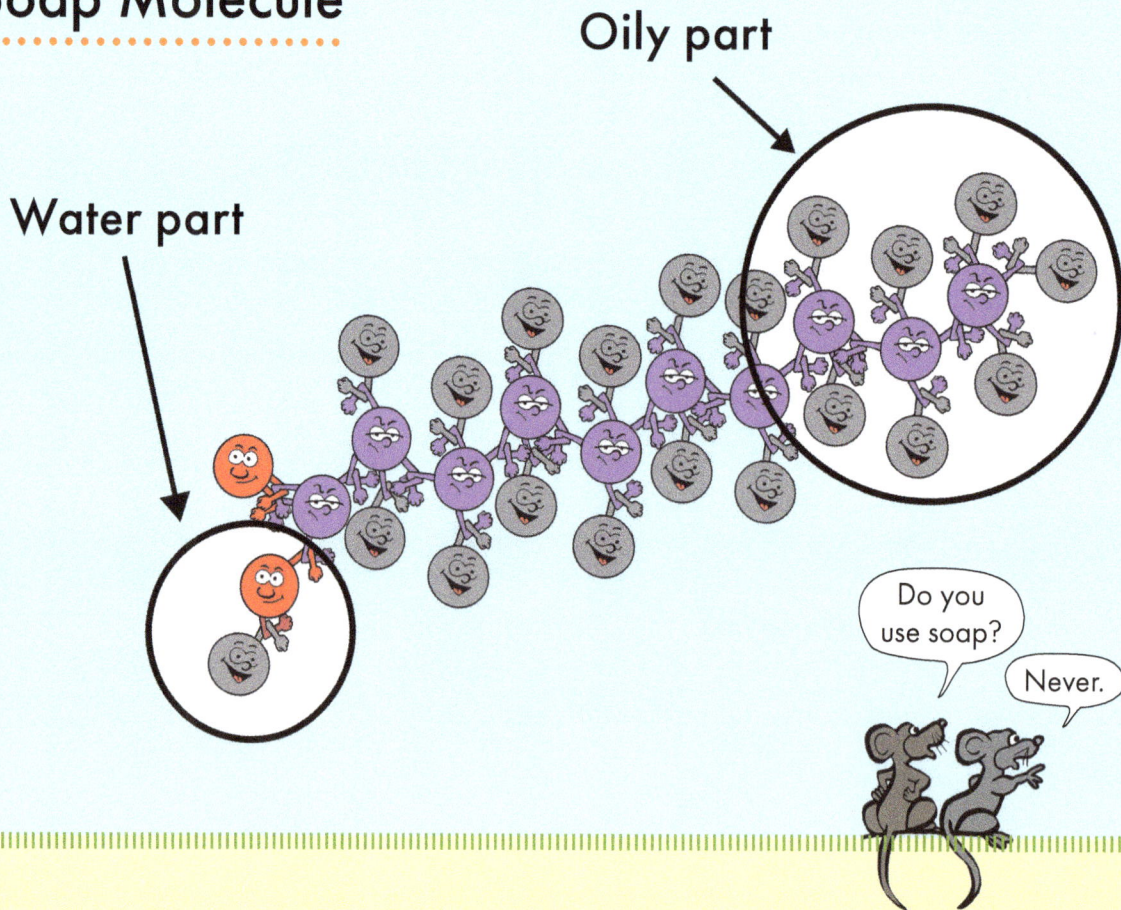

Oily part

Water part

Do you use soap?

Never.

Soap molecules surround oil molecules and form a ball.

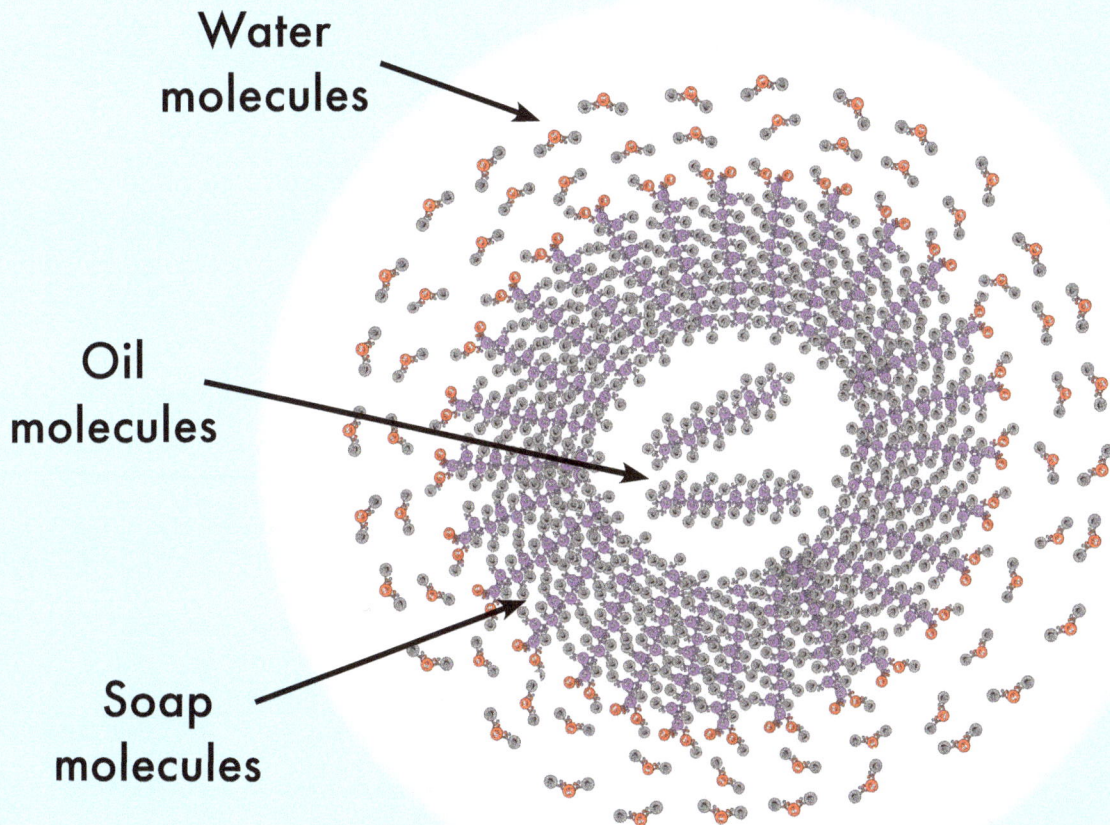

Water molecules

Oil molecules

Soap molecules

Inside the ball is the oily part of the soap molecule. This part traps the oil.

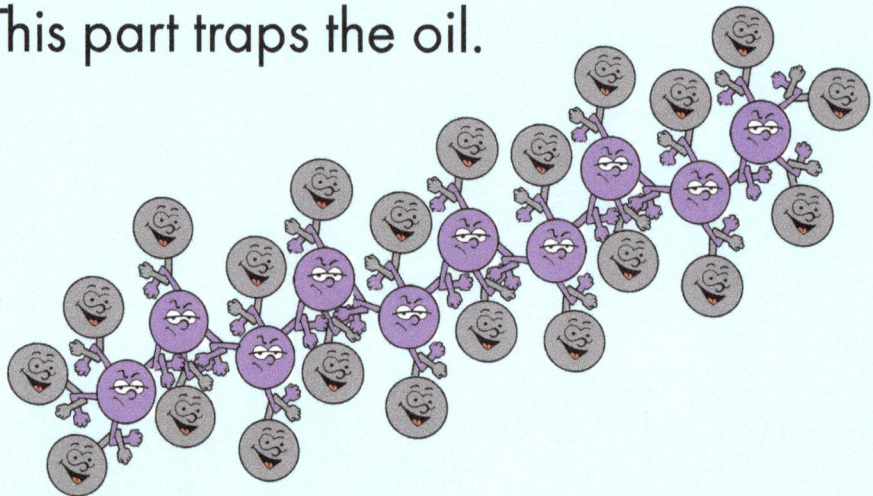

Outside the ball is the water part of the soap molecule. This part interacts with the water.

Water can wash the oily soap ball away.

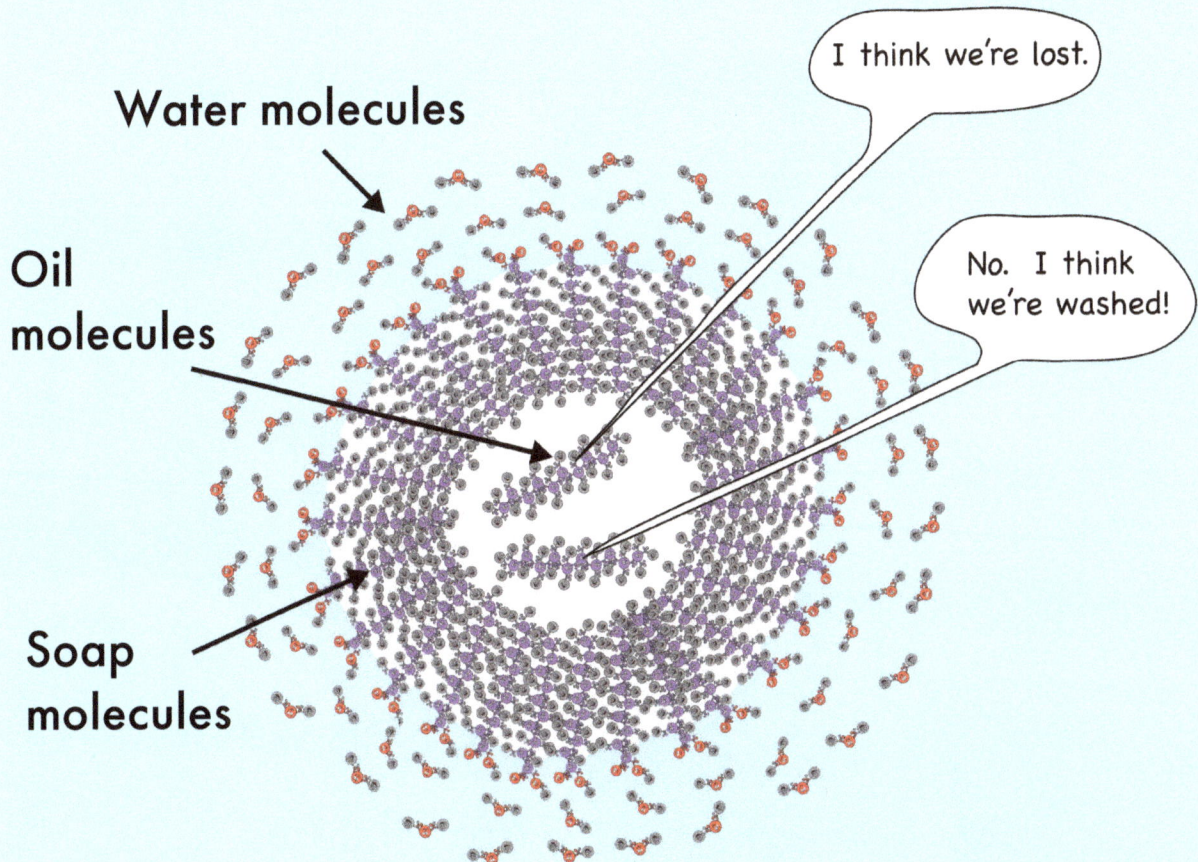

Water molecules

Oil
molecules

Soap
molecules

I think we're lost.

No. I think
we're washed!

Mixtures are found everywhere!

The foods we eat are mixtures.

The clothes and jewelry we wear are mixtures.

The fuels we burn are mixtures.

Mixtures help us stay warm, feed our bodies, and improve our lives.

How to say science words

atom (AA-tum)

evaporation (i-vap-uh-RAY-shuhn)

filtration (fil-TRAY-shuhn)

mixture (MIKS-chur)

molecule (MAH-lih-kyool)

paper chromatography

(PAY-puhr kroh-muh-TAH-gruh-fee)

www.ingramcontent.com/pod-product-compliance
Lightning Source LLC
Chambersburg PA
CBHW040150200326
41520CB00028B/7557